An Introduction to Stabilized Soil Construction Control

J. Paul Guyer, P.E., R.A.

Editor

The Clubhouse Press
El Macero, California

CONTENTS

1. GENERAL

2. COMPACTED SOIL-CEMENT

3. COMPACTED SOIL-LIME

4. CHANGES IN SOIL PROPERTIES

(This publication is adapted from the *Unified Facilities Criteria* and other resources of the United States government which are in the public domain, have been authorized for unlimited distribution, and are not copyrighted.)

(Figures, tables and formulas in this publication may at times be a little difficult to read, but they are the best available. **DO NOT PURCHASE THIS PUBLICATION IF THIS LIMITATION IS UNACCEPTABLE TO YOU.**)

1. GENERAL. Soil stabilization is the chemical or mechanical treatment of soil to improve its engineering properties. Chemically stabilized soils consist of soil and a small amount of additive such as cement, fly-ash, or lime. The additive is mixed with the soil, and the mixture is used in compacted fills, linings, or blankets. Quality and uniformity of the admixture and the uniformity of moisture are closely controlled to produce a high quality end product. Therefore, processing equipment and procedures are specified to ensure that the relatively small amount of additive is uniformly distributed throughout the soil mixture before placement and compaction. Uniformity of soil in the mixture is a major factor in controlling desired uniformity of the final product; and soil gradation, plasticity, and moisture content should be controlled prior to mixing with the additive or prior to adding the stabilizer for in-place mixing. A mixing plant must be calibrated over the range of soil gradation stated in the specifications. Adjustments to the mixing plant should then be made during construction to accommodate variation in soil gradation or for other variable conditions. These adjustments are based on mixing plant calibrations obtained before and during construction. Sometimes, soils are stabilized to deeper depths by grouting or by injection methods to solve particular foundation problems. These methods for stabilization require specialized knowledge and understanding of the materials being used, and quality control procedures are specifically developed for the particular situation.

2. COMPACTED SOIL-CEMENT.

2.1 DESIGN CONSIDERATIONS. Compacted soil-cement has been used as upstream slope protection for embankment dams and as blankets, linings, or other applications. Soilcement used as upstream slope protection is placed in successive horizontal layers ranging from 150 to 300 mm (6 to 12 in) in compacted thickness to protect the slope from wave action. The layers are placed successively up the slope, and the outer edges form a stair-step pattern. When soil-cement is placed as a blanket or lining it is usually placed in layers up to 600 mm (2 ft) thick with the layers parallel to the slope. Although roads may be a minor part of work, compacted soil-cement has been used extensively for construction of road bases by others. Figures 3-42 and 3-43 show soil-cement facings constructed on the upstream slope of two structures. To satisfy design requirements for slope protection, a layer must be:

- Formed into a homogeneous, dense, permanently cemented mass that fulfills the requirements for compressive strength
- In intimate contact with earth slopes, abutments, or concrete structures

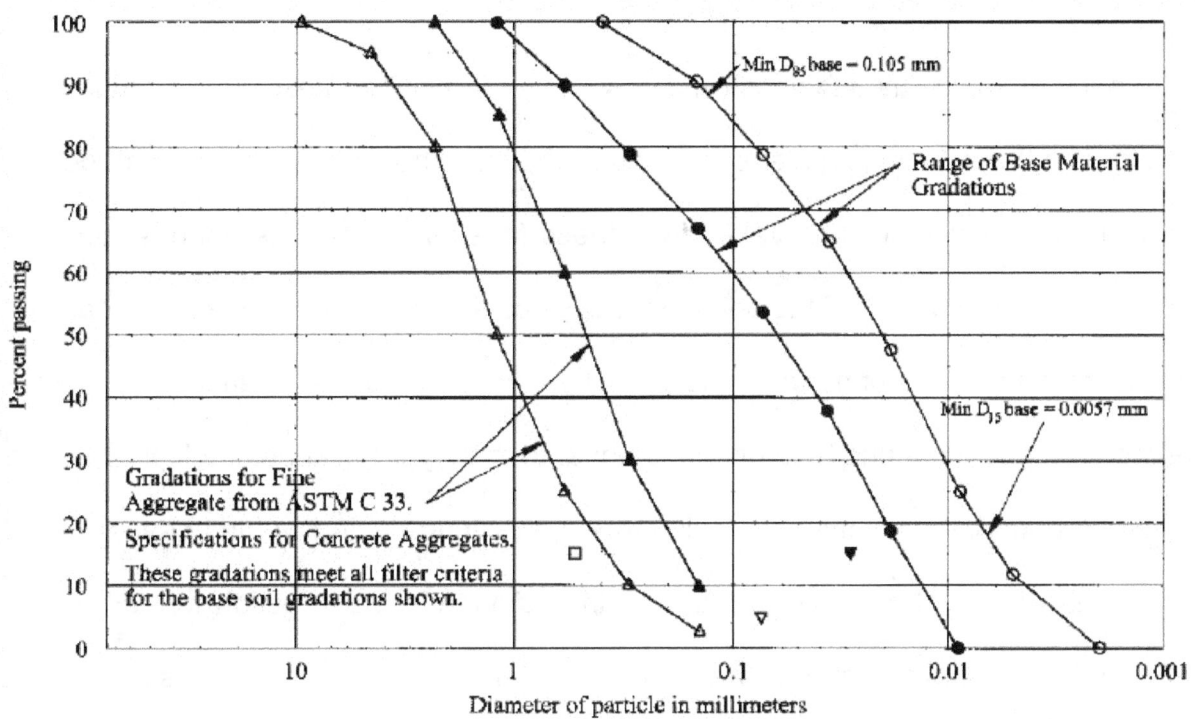

Figure 3-41

Filter gradations.

- **DURABLE AND RESISTANT** to "wetting and drying" and "freezing and thawing" action of water

- **STABLE WITH RESPECT TO** the structure and of sufficient thickness (mass) to resist displacement and uplift

Performance of soil-cement facings on dams has generally been excellent. However, inspections of these facings have revealed that the bond between lifts is a weak point in the facing. Test results on cores taken from these faces show that the bond is much weaker than the remainder of the soil-cement. Since the layers are not well bonded, they perform as a series of nearly horizontal slabs on the slope of a dam. Each slab is offset from the previous one by a distance equal to the layer thickness multiplied by the slope of the dam. If the layers were well bonded, the entire facing would act as a massive unit instead of as individual layers, and damage by wave action would be greatly minimized. Several studies have been performed to identify methods for enhancing bond between layers. Currently, the most promising method investigated is to apply a water-cement slurry to a layer just before placing the overlying layer. This technique was used at Davis Creek Dam in Nebraska to improve bonding between layers and to improve overall durability of the facing for an elevation of about 6 m (20 ft) at normal reservoir operating level.

Figure 3-42

Soil-cement facing at Lubbock Regulating Reservoir in Texas.

2.2 CONSTRUCTION PROVISIONS. The specifications describe the type and amount of cement, the quality and amount of water, and the borrow area for soil or aggregate. The permissible range of soil and aggregate gradation also is specified. If investigations show the deposit is variable, selective excavation and processing may be required to produce uniformity. Oversize particles and other objectionable materials must be removed. A stationary mixing plant is usually required. Either a batch type or a continuous-feed pugmill type plant is acceptable. Control over mixing time; positive interlocking of cement and soil flow; and controls for accurately proportioning soil, cement, and water all from appropriate storage should be incorporated into the plant design. A mixing plant used at Lubbock Regulating Reservoir in west Texas is shown on figure 3-44. Trucks for transporting the soil-cement mixture should have tight, clean, smooth beds and protective covers. The soilcement spreader used for laying the soil-cement must produce a smooth uniform loose layer of required width and thickness. Usually, layers are placed horizontally; however, a slope toward the outer edge as steep as 8h:1v is sometimes permitted to increase working width. The maximum time for hauling and spreading the soil-cement after mixing is usually specified as 30 minutes. The tractor and spreader box used for placing soil-cement on Merritt Dam and Reservoir in Nebraska is shown on figure 3-45. Generally, compaction must be completed within 60 minutes after spreading with no more than 30 minutes between operations. Compaction is accomplished by several passes of a sheepsfoot tamping roller, followed by several passes of a pneumatic-tire roller, as shown on figure 3-46. The minimum number of passes by each roller should be determined by constructing a test section. The rollers should have provisions for ballast loading so the masses can be adjusted to provide optimum compaction. Vibratory, smooth steel drum rollers have been used to compact coarse grained soil-cement. A combination of vibratory and static mode may be used. The roller mass and the frequency and amplitude of vibration should be set for optimum compaction without damage to the soil-cement and to minimize localized shear failure in the upper part of a layer.

Figure 3-43

Soil-cement facing at Merritt Dam in Nebraska.

A test section should always be constructed to determine: optimum equipment usage, roller mass, and number of passes and vibratory characteristics if a vibratory roller is used. Rollers may be towed or self-propelled. After compaction, the compacted layer is cured by keeping the exposed surfaces continually moist using a fog spray until the overlying or adjacent layer is placed, or for a minimum of 7 days. A blanket of moist earth may be used for permanently exposed surfaces. The surface of a completed layer may require brushing to remove soil or other debris just before placing an overlying layer, as shown on figure 3-47.

2.3 CONTROL TECHNIQUES. Compacted soil-cement is similar to compacted earthwork in that careful observations and additional control tests are required during the early stages to check planned control against results obtained under field conditions. These observations and tests are used to establish placing conditions and to develop procedures for use in the remaining construction. Control should begin during excavation and stockpiling of the soil to ensure that the material is within gradation requirements and has a uniform moisture content. Gradation and moisture content are controlled by:

- Directing the excavation
- Mixing the stockpile by spreading and cross-dozing
- Sloping the stockpile surfaces to provide runoff without water catchments
- Sampling and testing during stockpiling

After the mixing plant is set up, the cement, soil, and water feeds must be calibrated individually to establish curves (or tables) of equipment settings versus quantity produced. The plant must be calibrated over the full range of anticipated production rates. A cement vane feeder and soil feed belt to a pugmill are shown on figure 3-48. Calibration is accomplished by timing and weighing quantities of moist soil, cement, and water to check feed and meter settings. To facilitate computation of dry soil production, moisture content of the wet soil should be determined at the plant by a quick method, such as the calcium carbide reaction device (USBR 5310), a microwave oven (USBR 5315), or the Moisture Teller device (USBR 5305).1 The cement feed is perhaps the most crucial calibration because of variation in the amount of cement can critically affect the properties of soil-cement. The cement feed is calibrated over the range of required cement content by varying the feeder speed. The cement feed is usually quite consistent because cement is a fairly uniform product and is fed under reasonably uniform conditions in a well-designed plant. Cement feed calibration allows accurate adjustment if soil feed rates change during progress of the job. Plant performance and production is checked under full load by:

Figure 3-44

Soil stockpile and pugmill for mixing soil-cement at Lubbock Regulating Reservoir in Texas.

- Timing and weighing truckloads of the soil-cement mixture

- Checking mixing time
- Inspecting the mixture for:
    - Uniformity in texture
    - Moisture
    - Distribution of cement

The soil and soil-cement mixture is sampled to determine:

- ù Cement content
- ù Moisture content by both quick and oven-dry methods
- ù Occurrence of "clay balls" (rounded balls of clayey fines and sand which do not break down during ordinary processing)

During construction, the inspector should check plant operation periodically. Inspection frequency will depend on performance of the plant and uniformity of soil gradation and moisture content. The soil feed rate should be checked at the beginning of each shift by timing and weighing a truckload of moist soil. By using a quick method to determine moisture content, the feed rate of dry soil can be computed, and the proper cement and water feed rates can be set. At regular intervals, the moisture content of the soil and soilcement mixture should be determined by a quick method to provide a basis for making adjustments in the water feed if necessary.

Figure 3-45

Soil-cement placement at Merritt Dam in Nebraska.

Also, the inspector does a check calibration each time a record control test is performed by timing the production and weighing a truckload of soil-cement. Based on water content and cement feed rate, the soil feed rate can be determined. The soil and soil-cement are sampled for moisture tests at the plant and for record tests in the laboratory. The placement inspector should observe placing and compaction procedures and should verify that loose lift thickness, texture, and surface are uniform and that the lift is to specified dimensions. Compaction begins with the roller starting at the outer edge. The number of passes and adequacy of overlap should be checked. When using a sheepsfoot roller, each pass over the entire width of the lift should be completed before the next pass begins, as the tamping feet tend to follow previous tracks if successive passes are made without overlap. Because of the sandy material, the sheepsfoot roller usually does not "walk out" completely but should begin to "walk out" on the last passes. Each layer is completed by compaction with a pneumatic-tire roller. This is in contrast to normal earthwork where successive layers are compacted together with a sheepsfoot roller. With the time limits used in soil-cement construction, it is usually not feasible to place successive layers quickly enough to compact them together; hence, each lift must be compacted individually. The pneumatic-tire roller also starts compaction at the outer edge of the lift to minimize the amount of lateral spreading of the soil-cement. Smooth, even surfaces after compaction are desired. Rutting usually indicates a high placement water content. Thickness and width of the completed compacted lift should be checked. The placement inspector should be sure that the surface of the compacted layer is kept continuously moist until the overlying layer is placed, or for a minimum of 7 days. If required, the surface should be thoroughly cleaned by brooming (brushing) just before placing an overlying layer (fig. 3-47). Access ramps for trucks hauling soil-cement are built of soil on the soil-cement facing; these should also be checked to verify that they satisfy specified thickness requirements [usually 600 mm (2 ft)]. Ramps are constructed to adequately protect the edge of the top layers on which soilcement is being placed. Inadequate ramps will result in damage to the outside edges of soil-cement layers and decrease durability of the finished product.

Figure 3-46

Compaction of soil-cement at Lubbock Regulating Reservoir in Texas.

2.4 CONTROL TESTING. During construction, control testing is required for every 500 m3 (500 yd3) of soil-cement placed, or a minimum of one test per shift. The control test consists of timing a sample through the process to ensure that specified time constraints are met. Times recorded are (1) when the truck leaves the plant, (2) when the load is spread, (3) when compaction is completed, and (4) when the control test is taken in the compacted material. Control testing begins by timing the production of a truckload of soil-cement and obtaining a representative sample of soil from the soil feed.

The mass of the timed truckload is determined, and a representative sample of soil-cement is obtained for laboratory testing. When the load is spread, the approximate center of the load is marked by the placement inspector. Moisture contents of the soil and soil-cement are determined by a quick method at the time of sampling, and the remainder of the samples are used for laboratory tests. For the soil, the percentage of fines is determined, and the moisture content is obtained by the standard ovendrying method. The cement content of the soil can be determined by chemical titration. If the untreated soil contains a significant amount of calcium, a chemical analysis of the soil is performed. The presence of calcium should have been determined when the soil was being stockpiled. A complete gradation and specific gravity should be determined on specimens from every fourth control test. For the soil-cement, the percentage of "clay balls" is determined, and the percentage of cement is determined by either the heat of neutralization method or the chemical titration method. A three-point compaction test is performed using the rapid method of construction control. The compaction test should be performed at the same time that material (from the sampled truck load) is being compacted. This is necessary to allow for time effects on compaction properties of soil-cement caused by hydration of cement. The remainder of the soil-cement sample is used to prepare three or four compression test specimens to be tested at ages 7, 28, and 90 days. One 7-day and two 28-day specimens should be formed for each control test with an additional specimen for a 90-day test for every fourth control test.

Figure 3-47

Power brooming of surface prior to placement of next lift at Lubbock Regulating Reservoir in Texas.

Specimens tested at 90-days are for design verification only and not for construction control. Specimens are placed at the density determined from the field density test of the compacted soil-cement layer. These specimens should be formed as soon as possible after the field density has been determined. Procedures for preparing specimens for compressive strength testing are discussed in USBR 5806 and testing of soil-cement cylinders is described in USBR 5810. A field density test is performed at the point marked by the placement inspector when the timed load was spread. This test should be performed as soon as compaction is complete. Care should be taken so the test is not performed where roller overlap has occurred. When the test hole is complete, but before determining the volume, the depth of the lift is measured and recorded. Control testing includes obtaining record cores from the compacted soil-cement at least 28, 90, and 360 days after completion. One hole should be drilled for each 5,000 m3 (5,000 yd^3) placed. Locations of core holes should be spaced to be representative of the area covered, with some cores near abutments or structures. In drilling, care should be exercised to obtain a continuous core; comments on bond strength should be included with other information accompanying the cores. The core holes should be carefully backfilled with cement grout and a reinforcing bar placed flush with the surface and located for future reference. Record cross sections of the compacted soil-cement are obtained at locations where 28-day core samples are taken. The compressive strength of a section of core from each of the holes should be determined in the field laboratory; this strength is compared to those of the record construction control cylinders representative of the immediate vicinity. The remainder of the cores should be sent to the laboratory for durability, direct shear, and compression tests. A summary of compressive strength and age of record core, and compressive strength of specimens of material in the immediate vicinity, should accompany the cores.

3. COMPACTED SOIL-LIME.

3.1 GENERAL. Use of lime as a soil additive is the oldest known method of chemical stabilization; it was used by the Romans to construct the Appian Way. Soil-lime is a mixture of soil (usually clay), lime, and water which is compacted to form a dense mass. Experience has shown that mixtures of most clay soils, either quick or hydrated lime, and water will form cementitious products in a fairly short period of time. Applications for water resources work have been limited to use of lime to stabilize expansive clay soils and dispersive clay soils.

3.2 ADDING LIME TO HIGHLY PLASTIC CLAY SOILS. Adding lime to highly plastic clay soils produces several effects on physical properties:

- The liquid limit decreases, and the plastic limit increases, radically decreasing the plasticity index (sometimes, by a factor of 4 or more)
- The finer clay-size particles agglomerate to form larger particles, which makes the soil more friable and easier to work. By absorbing water, lime also assists in breaking up clay clods during mixing.
- Lime dries the soil by absorbing water to hydrate the lime and makes wet soils easier to handle and compact.
- Unconfined compressive strength increases many fold.

Friant-Kern Canal, California, was constructed during 1945-51; about 87 km (54 mi) of the canal traverse an area of expansive clay. Of these 87 km, 37 km (23 mi) are earth lined, and the remaining 50 km (31 mi) are concrete lined. After 3 years of operation, portions of the canal traversing expansive clay soils began cracking, sloughing, and sliding with failures occurring in both the concrete-lined and earth-lined sections. Because these conditions caused continuing, expensive maintenance problems, in 1970, rehabilitation began for the worst failed areas. Lime stabilization was selected as the most effective method of treatment. Riprap that had been dumped into slide areas was removed. Then, all material to be stabilized with lime and recompacted was removed by a benching operation. A series of long sloping benches or ramps were cut from the top of the bank down to the canal bottom, with the cut extending far enough into the slope to remove the entire depth of required excavation material. Two-percent quicklime was spread over the bench surface, and 0.3 m (1 ft) of material from the bench was mixed with the quicklime, and the lime-clay mixture was pushed into the canal bottom. The material was spread on the canal bottom and an additional 2 percent lime was added. Water was added to at least 2 percentage points above optimum moisture, and about 0.3 m depth of material was mixed with dozers and graders. After about 2 m (6.6 ft) of material had been mixed and cured for 24 hours, dozers began spreading the material on the slopes, which were then compacted with a sheepsfoot roller moving up and down the slope. The side slopes were constructed in three compacted lifts for a 1.1 m (3.6 ft) compacted depth normal to the slope. In subsequent rehabilitation work, the lime-treated soil was placed and compacted in successive horizontal layers stepped up the slope in the same manner in which soil-cement is placed on the face of an earth embankment. This placement method was found much more desirable than placing material parallel to the slope. Placing and compaction were much more efficient, and the finished product was of higher quality. The amount of lime used was controlled by periodically placing a canvas on the ground where lime was to be spread, and after spreading, mass of lime was determined for a given area. Four-percent lime by dry mass of soil was used during rehabilitation of Friant-Kern Canal; compressive strength increased to 20 times that of untreated soil. The rehabilitation has proved durable after about 20 years of additional service.

3.3 ADDING LIME TO DISPERSIVE CLAY SOILS. Dispersive clay soils are those that erode in slow-moving or even quiet water by individual colloidal clay particles going into suspension and then being carried away by the flowing water. A concentrated leakage channel (crack) must be present for erosion to initiate in dispersive clay. This mechanism is totally different than that for piping where erosion begins at the discharge end of a leak and progresses upstream through the structure until it reaches the water source. The design lime content for controlling dispersive clay soils is generally defined as the minimum lime content required to make the soil nondispersive. In addition, it may be desirable to increase the shrinkage limit to near optimum water content to prevent cracking from drying when using lime-treated soil in surface layers. In all known cases investigated to date, dispersive clay soils were made nondispersive by addition of 1 to 4 percent lime by dry mass of soil. However, in construction specifications, the design lime content is often increased 0.5 to 1.0 percent to account for losses, uneven distribution, incomplete mixing, etc. Dispersive clay soils were identified throughout the borrow and foundation areas during investigations for McGee Creek Dam, Oklahoma; it was determined to be practical to stabilize these soils with lime during dam construction.

Since dispersive clay soils were not concentrated in specific areas, lime treatment was considered more economical than attempting to identify the randomly occurring dispersive clays and selectively wasting them. Treating these clay soils with lime rendered them nondispersive and allowed their use in constructing the embankment-foundation contacts as erosion resistant material on the downstream slope of the embankments, and for placement as specially compacted backfill in areas of high piping potential such as along conduits through the embankment. Material was used directly from the borrow areas to construct the remainder of the dam and dike embankments. Specially designed filters were used to guard against any possible erosion of dispersive clay soils in untreated areas. Another benefit of using lime-treated soil was the improved workability of some of the highly plastic clay soil encountered.

3.4 CONSTRUCTION PROCEDURES FOR LIME. TREATED DISPERSIVE CLAY SOILS. The following general procedures have proved satisfactory for handling, mixing, and placing lime-treated soil. Soil to be lime treated is pulverized in a high speed rotary mixer or with a disk harrow prior to applying lime, and the moisture content is brought to within 2 percent of optimum. Lime is uniformly spread on the pulverized soil to the specified percent lime by dry mass of soil. Lime is mixed with the soil using a rotary mixer, and additional water is added as necessary to again bring the mixture to within 2 percent of optimum (or other specified value). When mixing is completed, the soil-lime mixture is cured for at least 96 hours before placing and compacting. Exposed surfaces of the mixture are either lightly rolled to prevent moisture loss or the mixed material is stockpiled and the surface sealed. Each section of the foundation is carefully prepared coincident with final mixing and pulverization of the lime-treated material. The soil-lime is mixed until 100 percent passes the 25 mm sieve and 60 percent passes the 4.75-mm sieve (1 in and No. 4). Immediately after final mixing, the lime-treated earthfill is placed and compacted in horizontal lifts of no more than 150 mm (6 in) after compaction. The material is compacted to no less than 95-percent laboratory maximum dry density, using a tamping roller followed by a pneumatic-tire roller. The top of each compacted lift is scarified or disked before the next lift is placed [44]. The following items should be monitored to ensure high quality earthwork construction control:

- Soil pulverization (gradation)
- Lime content
- Soil dispersivity (before and after lime treatment)
- Compacted density
- Moisture content of both soil and soil-lime mixture

If embankment materials are placed on the compacted lime-treated earthfill within 36 hours, special curing provisions are not required. Otherwise, the exposed surface of the lime-treated earthfill is compacted with a pneumatic-tire roller to seal the surface, and it is sprinkled with water for 7 days or until embankment material is placed. Construction control testing is the same as that for other earthwork.

4. CHANGES IN SOIL PROPERTIES. Although soil is commonly considered a stable material, it is constantly changing, either gradually from solid rock to increasingly finer particles or, conversely, gradually changing back to rock. In most soils, this change is sufficiently gradual that it is not a concern. However, in some soils, the change is rapid enough to be important in the life of an engineered structure. Soils where change may be important include those with appreciable quantities of: (1) organic matter, (2) soluble solids, or (3) minerals of volcanic origin. Residual soils may be in a state of chemical alteration such that during placement, they will have one set of characteristics; later, during the life of the hydraulic structure, they may have very different characteristics. Frequently, existing soil deposits in their natural state have been stable for many years and give every indication that they will remain so. Nevertheless, human changes may result in failure of some soils. One of the soils prone to failure is called sensitive clay. This type of clay in an undisturbed condition has substantial shear strength, which to a large extent is lost upon being remolded. Very loose, saturated, fine sand and silt when subjected to dynamic loading such as an earthquake or vibration from machinery will lose strength and behave like a viscous liquid. This phenomenon is known as liquefaction. Another group of soils exists where minor changes in moisture content result in an abrupt change in shear strength. In some cases, these soils, such as loessial soils, have been deposited in a very loose state and exhibit change in shear strength and can collapse and subside when the moisture content is increased. Swelling clays frequently exhibit a change in strength characteristics caused by an increase in moisture. Among the soils that, through desiccation, consolidation, and chemical action, have changed to forms commonly regarded as rock are varieties of shale, sandstone, and limestone. When these rocks are exposed to air, marked changes in characteristics can occur. Some shales flake off, air slake, or weather rapidly turning into soil. Some shales may dry out without any apparent effect, but if rewetted, they deteriorate rapidly into very soft clay. Some sandstones and limestones harden on exposure to air and retain their improved qualities, while other limestones and sandstones break down rapidly with fluctuating temperature and moisture content. Although deterioration is rare, if unrecognized, failure can occur without advance warning. Engineering practices for treating soils that deteriorate are not well known. Where situations as described above are suspected, the situation should be reviewed by specialists in this field, and specialized treatment may be necessary.

4.1 WORKABILITY. Although laboratory testing indicates the maximum extent to which engineering properties such as shear strength, volume change, and permeability of a given soil may vary, achieving these limits in engineering practice is seldom practical. The ease with which satisfactory values of engineering properties can be economically reached is an important attribute of a soil, a soil deposit, or a foundation. The cost to procure a unit volume of soil and place it (as in a structure) or for treating a unit amount of foundation varies widely, not only according to soil type but according to type and size of structure. Also, cost is influenced by the kind of equipment available and by current available labor. If a project is sufficiently large so that special equipment can be economically used, maximum efficiency in construction is most likely to be achieved. However, the soils selected must be workable by such methods. Where separation of oversize is not required and where mixing requirements are minimized, borrow pits which can be preprocessed to optimum moisture content are preferable and usually more economical, even though longer haul distances may be required for their effective use. In practice, situations arise where separation of oversize is economical; also, cases exist where mixing two varieties of soil is worthwhile. Instances occur where extensive efforts to obtain maximum moisture control are justified; however, such operations should be avoided if possible. Test procedures do not exist for measuring workability. Rather, all pertinent information concerning a soil, a borrow pit, or a foundation is tabulated so the various design possibilities can be evaluated. An engineering use chart provides qualitative information on the workability of soils as a construction material and the relative desirability of various soil types according to structure. Borrow pits may be evaluated according to amount of work required. Because equipment mobilization is charged against a soil deposit, unit cost decreases appreciably as the volume of work increases. The change in unit cost for excavation up to about 100,000 m^3 (100,000 yd^3) is noticeable; then to 1 million m^3 (1 million yd^3) it is gradual; and beyond that range, unit cost is nearly constant. Transportation costs are nearly constant above about 100,000 m^3 (100,000 yd^3), depending only on distance. Moisture control costs depend primarily on the uniformity and slope of the borrow area and the availability of water. Excavation costs are influenced somewhat by topography of the borrow area. Borrow pits slightly higher in elevation than the work structure are preferable to those below the work.

4.2 FROST ACTION. Heaving of subgrades caused by formation of ice lenses and subsequent loss of shear strength upon thawing is known as frost action. Water rises by capillarity and by thermal gradient toward the freezing zone and forms lenses of ice, which heave the soil. Soils most susceptible to frost action are those in which capillarity can develop but are sufficiently pervious to allow adequate water movement upward from below the freezing zone. Freezing of the pore water in saturated fine-grained soils, called closed-system freezing, decreases the density of soil by expansion but does not result in appreciable frost heave unless water movement can take place from below. The severity of frost heave depends on three factors: (1) type of soil, (2) availability of free water, and (3) time rate of fluctuation of temperature about the freezing point. Soils having a high percentage of silt-size particles are the most frost susceptible. Such soils have a network of small pores that promote migration of water to the freezing zone. Silt (ML, MH), silty sand (SM), and clays of low plasticity (CL, CL-ML) are in this category. Tests from 1950 to 1970 form the basis for figure 1-34; it relates frost susceptibility in terms of average rate of heave in percent by mass finer than the U.S.A. Standard 75-µm (No. 200) sieve. The figure shows most soil types have a wide range of frost susceptibility without a sharp dividing line between frost-susceptible and nonfrost-susceptible soils. Nevertheless, silts, clayey silts, and silty sands have the highest potential for frost heave followed by gravelly and sandy clay, clayey sand, and clayey gravel. Soils with the lowest potential to heave are sandy gravels, clean sands, and silty sands with less than 3 percent finer than the 75-µm (No. 200) sieve. In the absence of a source of free water, frost heave is limited to the increase in volume due to freezing of pore water. This upper limit amounts to about 9 percent of pore water volume.

4.3 ERODIBILITY. Erosion has been defined as ". . . a process of detachment and transport of soil particles or particle groups by the forces of water, wind, ice, and gravity." Erodibility is the susceptibility of a soil to erode. The processes that influence erosion of cohesionless soil particles have been understood for many years. A cohesionless soil particle resting on the side or bottom of a stream or canal is acted on by gravity and by a tractive or erosive force caused by movement of water past the particle. Thus, erosion resistance of cohesionless soils depends on the applied tractive force and the mass of the particle expressed in terms of mean particle diameter (D50). Figure 1-35 presents data collected by a number of investigators showing the relationship between critical tractive force, the force required to start erosion, and the mean particle diameter. The processes that influence erosion of cohesive soils have been studied for a number of years but still are not completely understood. Over the years, investigators have attempted to correlate tractive force to various parameters and properties of cohesive soil. Strong, consistent correlation has not been found. Field performance data have been collected on operating canals and streams as well as laboratory data collected from various erosion devices, including flumes, erosion tanks, submerged hydraulic jets, and rotating cylinders. These studies have helped identify the properties that influence erosion of cohesive soils but have not provided quantifiable correlations between laboratory tests and erodibility in the field. Some soil and fluid properties that may influence the erosion process in cohesive soil include:

- dry density and moisture content,
- grain-size distribution,
- Atterberg limits,

- undrained shear strength,
- clay mineralogy,
- pore-water chemistry,
- stress history,
- soil structure and fabric,
- chemistry of the eroding fluid,
- temperature,
- viscosity of the eroding fluid, and
- applied tractive stress.

4.4 DISPERSIVE CLAY. Some unique cohesive soils have been found to be highly erodible. These soils are designated dispersive clay because they erode when the individual clay particles disperse (go into suspension) even in the presence of still water. Dispersive clays cannot be distinguished from nondispersive clays by conventional index tests such as gradation, Atterberg limits, or compaction characteristics. Dispersive characteristics are determined by performing three standardized tests on the questionable clay sample material. The three test results are combined to rate the clay as dispersive, intermediate, or nondispersive. Chemical tests to determine quantity and type of dissolved salts in the pore water are also useful in determining dispersive potential of clay soils. Dispersive clays can be made nondispersive by adding a small percentage of hydrated lime (about 2 to 4 percent by dry mass of soil) to the soil. Detrimental effects of dispersive clay (in hydraulic structures) can also be minimized by proper zoning and by using designed granular filters to prevent piping failures.

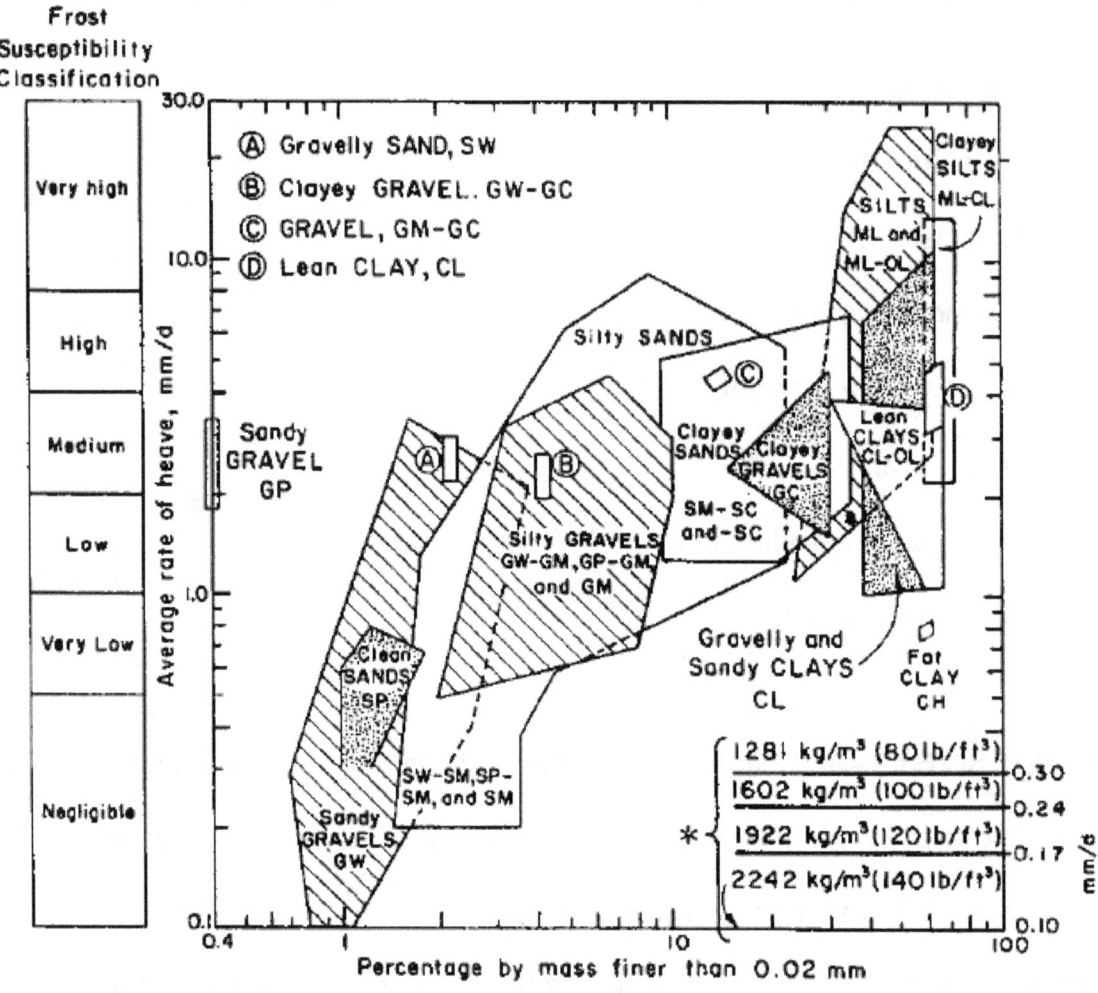

Figure 1-34

Frost susceptibility classification by percentage of mass finer than 0.02 mm

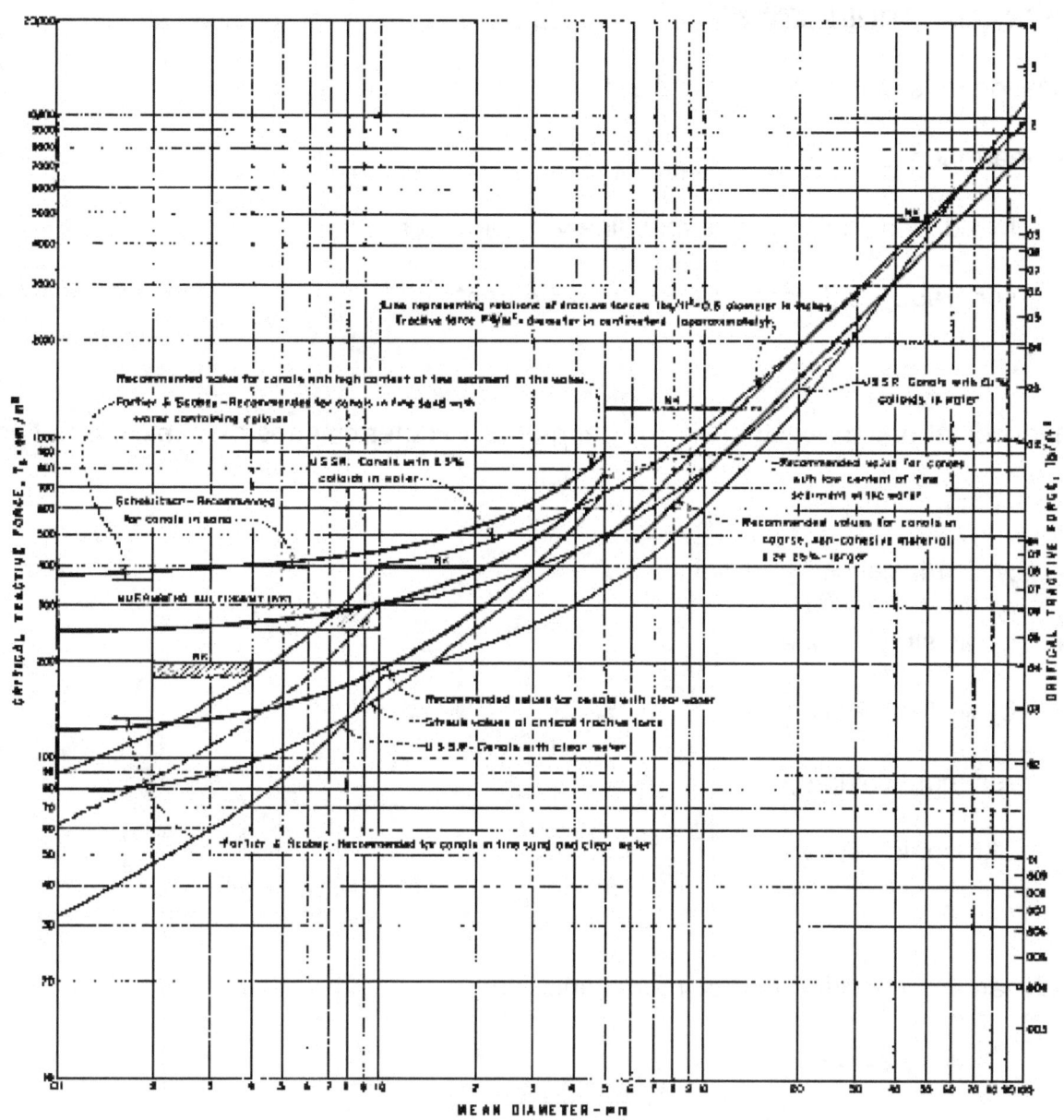

Figure 1-35

Limiting tractive forces recommended for canals and observed in rivers [

4.5 DYNAMIC PROPERTIES. The response of soil to cyclic or dynamic stress application must be considered in the design of structures subjected to:

- earthquake loading,
- foundations subjected to machine vibrations, and
- subgrades and base courses for pavement.

A variety of laboratory equipment has been used to determine the dynamic properties of soil including:

- cyclic triaxial compression,
- cyclic simple shear,
- cyclic torsional shear,
- resonant column,
- ultrasonic devices.

The dynamic properties of most interest include:

- shear modulus,
- damping ratio,
- dynamic shear strength,
- pore-pressure response.

Detailed dynamic properties test procedures are not included in this discussion.

www.ingramcontent.com/pod-product-compliance
Lightning Source LLC
Chambersburg PA
CBHW080855170526
45158CB00009B/2746